INTERNATIONAL CENTRE FOR MECHANICAL SCIENCES

GYULA KATONA
MATHEMATICAL INSTITUTE
HUNGARIAN ACADEMY OF SCIENCES, BUDAPEST

GENERAL THEORY OF NOISELESS CHANNELS

LECTURES HELD AT THE DEPARTMENT
FOR AUTOMATION AND INFORMATION
JUNE 1970

UDINE 1970

COURSES AND LECTURES - No. 31

ISBN 978-3-211-81167-2 ISBN 978-3-7091-2872-5 (eBook)
DOI 10.1007/978-3-7091-2872-5

Originally published by Springer Vienna in 1970

P R E F A C E

"Noiseless channels" is an expression like "rose without a thorn". In practical cases we have almost always noisy channels. However, it is useful to examine the noiseless channels because there are channels with small noise, we may consider them to be noiseless channels.

On the other hand, studying noiseless channels we can get directions about the properties of noisy channels, which are more complicated, thus it is much more difficult to study them directly.

This short survey paper is the written form of my 8 lectures organized by CISM in June 1970.

In the lecture notes only the elementary probability theory is used and some elementary properties of the information-type functions. These properties are proved in the "Preliminaries" written by Professor I. Csiszár.

I would like to express my thanks to the CISM for giving me this opportunity.

I hope this paper will be an almost noiseless channel from the author to the readers.

Udine, June 22, 1970. G. Katona

Preliminaries

In this section we summarize some basic definitions and relations which will be used freely in the sequel : the simple proofs will be sketched only.

The term "random variable" will be abbreviated as RV ; for the sake of simplicity, attention will be restricted to the case of discrete RV's, i.e., to RV's with values in a finite or countably infinite set.

ξ, η, ζ will denote RV's with values in the (finite or countably infinite) sets X, Y, Z.

All random variables considered at the same time will be assumed to be defined on the same probability space. Recall that a probability space is a triplet $(\Omega, \mathcal{F}, P)$ where Ω is a set (the set of all conceivable outcomes of an experiment), $\mathcal{F}$ is a σ-algebra of subsets of Ω (the class of observable events) and P is a measure (non-negative countably additive set function) defined on $\mathcal{F}$ such that $P(\Omega)=1$. Rv's ξ, η etc. are functions $\xi(\omega)$, $\eta(\omega)$ etc. $(\omega \in \Omega)$. The probability $P\{\xi = x\}$ is the measure of the set of those ω's for which $\xi(\omega)=x$; similarly, $P\{\xi=x, \eta=y\}$ is the measure of the set of those ω's for which $\xi(\omega)=x$ and $\eta(\omega)=y$.

The conditional probability $P\{\xi = x \mid \eta = y\}$ is defined as $\frac{P\{\xi = x, \eta = y\}}{P\{\eta = y\}}$ (if $P\{\eta=y\}=0$, $P\{\xi=x \mid \eta=y\}$ is undefined).

Definition 1. The RV's defined by

(1) $\iota_\xi = -\log_2 P\{\xi = x\}$ if $\xi = x$

(2) $\iota_{\xi\wedge\eta} = \log_2 \dfrac{P\{\xi = x, \eta = y\}}{P\{\xi = x\} P\{\eta = y\}}$ if $\xi = x, \eta = y$

are called the entropy density of ξ and the information density of ξ and η, respectively.

(3) $\iota_{\xi|\eta} = -\log_2 P\{\xi = x \mid \eta = y\}$ if $\xi = x, \eta = y$

(4) $\iota_{\xi|\eta,\zeta} = -\log_2 P\{\xi = x \mid \eta = y, \zeta = z\}$ if $\xi = x, \eta = y, \zeta = z$

are conditional entropy densities and

(5) $\iota_{\xi\wedge\eta|\zeta} = \log_2 \dfrac{P\{\xi = x, \eta = y \mid \zeta = z\}}{P\{\xi = x \mid \zeta = z\} P\{\eta = y \mid \zeta = z\}}$ if $\xi=x, \eta=y, \zeta=z$

is a conditional information density.

Remark. Entropy density is often called "self-information" and information density "mutual information". In our terminology, the latter term will mean the expectation of $\iota_{\xi\wedge\eta}$.

Definition 2. The quantities

$$E(\xi) \overset{\text{def}}{=} E\iota_{\xi} = -\sum_{x\in X} P\{\xi=x\} \log_2 P\{\xi=x\} \tag{6}$$

$$I(\xi\wedge\eta) \overset{\text{def}}{=} E\iota_{\xi\wedge\eta} = \sum_{x\in X, y\in Y} P\{\xi=x,\eta=y\} \log_2 \frac{P\{\xi=x,\eta=y\}}{P\{\xi=x\}P\{\eta=y\}} \tag{7}$$

are called the entropy of ξ and the mutual information of ξ and η, respectively.

The quantities

$$H(\xi|\eta) \overset{\text{def}}{=} E\iota_{\xi|\eta} = -\sum_{x\in X, y\in Y} P\{\xi=x,\eta=y\} \log_2 P\{\xi=x|\eta=y\} \tag{8}$$

$$H(\xi|\eta,\zeta) \overset{\text{def}}{=} E\iota_{\xi|\eta,\zeta} = -\sum_{x\in X, y\in Y, z\in Z} P\{\xi=x,\eta=y,\zeta=z\} \log_2 P\{\xi=x|\eta=y,\zeta=z\} \tag{9}$$

are called conditional entropies and

$$I(\xi\wedge\eta|\zeta) \overset{\text{def}}{=} E\iota_{\xi\wedge\eta|\zeta} = \sum_{x\in X, y\in Y, z\in Z} P\{\xi=x,\eta=y,\zeta=z\} \log_2 \frac{P\{\xi=x,\eta=y|\zeta=z\}}{P\{\xi=x|\zeta=z\}P\{\eta=y|\zeta=z\}} \tag{10}$$

is called conditional mutual information.

Here terms like $0 \log_2 0$ or $0 \log_2 \frac{0}{0}$ are meant to be 0.

The quantities (6)-(10) are always non-negative (for (7) and (10) this requires proof ; see (17), (18)) but they may be infinite. The latter contingency should be kept in mind ; in particular, identities like $I(\xi\wedge\eta) = H(\xi) - H(\xi|\eta)$ (cf. (21)) are valid only under the condition that they do not contain the undefined expression $+\infty - \infty$.

$H(\xi)$ is interpreted as the measure of the average amount of information contained in specifying a particular value of ξ; $I(\xi \wedge \eta)$ is a measure of the average amount of information obtained with respect to the value of η when specifying a particular value of ξ. Conditional entropy and conditional mutual information are interpreted similarly. Logarithms to the basis 2 (rather than natural logarithms) are used to ensure that the amount of information provided by a binary digit (more exactly, by a random variable taking on the values 0 and 1 with probabilities 1/2) be unity. This unit of the amount of information is called <u>bit</u>.

The interpretation of the quantities (6)-(10) as measures of the amount of information is not merely a matter of convention ; rather, it is convincingly suggested by a number of theorems of information theory as well as by the great efficency of heuristic reasonings based on this interpretation. There is much less evidence for a similar interpretation of the entropy and information densities. Thus we do not insist on attaching any intuitive meaning to the latters ; they will be used simply as convenient mathematical tools.

A <u>probability distribution</u>, to be abbreviated as PD, on the set X is a non-negative valued function $p(x)$ on X with $\sum_{x \in X} p(x) = 1$; PD's will be denoted by script letters, e.g. $\mathcal{P} = \{ p(x),\ x \in X \}$.

<u>Definition 3</u>. The I-divergence of two PD's $\mathcal{P}=\{p(x),\ x\in X\}$ and $\mathcal{Q}=\{q(x), x\in X\}$ is defined as

$$I(\mathcal{P}\|\mathcal{Q}) = \sum_{x\in X} p(x)\ \log_2 \frac{p(x)}{q(x)}\ . \tag{11}$$

Here terms of the form $a\log_2\frac{a}{0}$ with $a>0$ are meant to be $+\infty$.

<u>Lemma 1</u>. Using the notations $p(A)=\sum_{x\in A} p(x)$, $q(A)=\sum_{x\in A} q(x)$ we have for an arbitrary subset A of X

$$\sum_{x\in A} p(x)\ \log_2\frac{p(x)}{q(x)} \geq p(A)\ \log_2\frac{p(A)}{q(A)}\ ; \tag{12}$$

if $q(A)>0$ the equality holds iff*) $p(x)=\frac{p(A)}{q(A)}\,q(x)$ for every $x\in A$. In particular, setting $A=X$:

$$I(\mathcal{P}\|\mathcal{Q})\geq 0\ , \qquad \text{equality iff}\quad \mathcal{P}=\mathcal{Q}. \tag{13}$$

<u>Proof</u>. The concavity of the function $f(t)= = \ln t$ implies $\ln t\leq t-1$, with equality iff $t=1$. Setting now $t=\frac{q(x)}{p(x)}\frac{p(A)}{q(A)}$ one gets $\ln\frac{q(x)}{p(x)}\leq \ln\frac{q(A)}{p(A)}+\frac{q(x)}{p(x)}\frac{p(A)}{q(A)}-1$ whenever $p(x)\,q(x)>0$, with equality iff $\frac{q(x)}{p(x)}=\frac{q(A)}{p(A)}$. Multiplying by $p(x)$ and summing for every $x\in A$ with $p(x)>0$ (one may obviously assume that then $q(x)>0$ too) (12)follows, including the condition for equality. The choice of the basis of the logarithms being clearly immaterial. The I-divergence $I(\mathcal{P}\|\mathcal{Q})$ is a measure of how different the PD $\mathcal{P}$ is from the PD $\mathcal{Q}$ (however note, that in general $I(\mathcal{P}\|\mathcal{Q}) = I(\mathcal{Q}\|\mathcal{P})$). If $\mathcal{P}$ and $\mathcal{Q}$ are two

*)Iff is an abbreviation for "if and only if".

hypothetical PD's on X then $I(\mathcal{P}\|\mathcal{Q})$ may be interpreted as the average amount of information in favour of $\mathcal{P}$ and against $\mathcal{Q}$, obtained from observing a randomly chosen element of X, provided that the PD $\mathcal{P}$ is the true one.

The <u>distribution of a RV</u> ξ is the PD $\mathcal{P}_\xi$ defined by

(14) $$\mathcal{P}_\xi = \{p_\xi(x),\ x\in X\},\quad p_\xi(x) = P\{\xi = x\}.$$

The <u>joint distribution</u> $\mathcal{P}_{\xi\eta}$ of the RV's ξ and η is defined as the distribution of the RV (ξ,η) taking values in $X\times Y$ i.e. $\mathcal{P}_{\xi\eta} = \{p_{\xi\eta}(x,y),\ x\in X, y\in Y\}$, $p_{\xi\eta}(x,y) = P\{\xi = x, \eta = y\}$.

From (7) and (11) it follows

(15) $$I(\xi\wedge\eta) = I(\eta\wedge\xi) = I(\mathcal{P}_{\xi\eta}\|\mathcal{P}_\xi\times\mathcal{P}_\eta)$$

where $\mathcal{P}_\xi\times\mathcal{P}_\eta = \{p_\xi(x)\,p_\eta(y),\ x\in X, y\in Y\}$ and also

(16) $$I(\xi\wedge\eta) = \sum_{x\in X} p_\xi(x)\, I(\mathcal{P}_{\eta|\xi=x}\|\mathcal{P}_\eta)$$

where $\mathcal{P}_{\eta|\xi=x} = \{p_x(y),\ y\in Y\}$, $p_x(y) = P\{\eta = y\ \xi = x\}$.
(15) and (13) yield

(17) $I(\xi\wedge\eta)\geq 0$, equality iff ξ and η are independent.

By a comparison of (7) and (10), this implies

(18) $I(\xi\wedge\eta\,|\,\zeta)\geq 0$, equality iff ξ and η are condition

ally independent for ζ given.

Let us agree to write $\iota_{\xi,\eta}$ for $\iota_{(\xi,\eta)}$ (entropy density of the RV (ξ,η)), $\iota_{\xi,\eta\wedge\zeta}$ for $\iota_{(\xi,\eta)\wedge\zeta}$ (information density of the RV's (ξ,η) and ζ) etc. ; omitting the brackets will cause no ambiguities.

<u>Theorem 1</u>. (Basic identities)

$$\iota_{\xi,\eta} = \iota_{\xi|\eta} + \iota_{\eta} \qquad H(\xi,\eta) = H(\xi|\eta) + H(\eta) \tag{19}$$

$$\iota_{\xi,\eta|\zeta} = \iota_{\xi|\eta,\zeta} + \iota_{\eta|\zeta} \qquad H(\xi,\eta|\zeta) = H(\xi|\eta,\zeta) + H(\eta|\zeta) \tag{20}$$

$$\iota_{\xi} = \iota_{\xi|\eta} + \iota_{\xi\wedge\eta} \qquad H(\xi) = H(\xi|\eta) + I(\xi\wedge\eta) \tag{21}$$

$$\iota_{\xi|\zeta} = \iota_{\xi|\eta,\zeta} + \iota_{\xi\wedge\eta|\zeta} \qquad H(\xi|\zeta) = H(\xi|\eta,\zeta) + I(\xi\wedge\eta|\zeta) \tag{22}$$

$$\iota_{\xi_1,\xi_2\wedge\eta} = \iota_{\xi_1\wedge\eta} + \iota_{\xi_2\wedge\eta|\xi_1} \; ; \; I(\xi_1,\xi_2\wedge\eta) = I(\xi_1\wedge\eta) + I(\xi_2\wedge\eta|\xi_1) \tag{23}$$

$$\iota_{\xi_1,\xi_2\wedge\eta|\zeta} = \iota_{\xi_1\wedge\eta|\zeta} + \iota_{\xi_2\wedge\eta|\xi_1,\zeta} \; ; \; I(\xi_1,\xi_2\wedge\eta|\zeta) = I(\xi_1\wedge\eta|\zeta) + I(\xi_2\wedge\eta\ \xi_1,\zeta) \tag{24}$$

<u>Proof</u>. Immediate from definitions 1 and 2

<u>Theorem 2.</u> (Basic inequalities)

The information quantities (6)-(10) are non-negative ;

$$H(\xi,\eta) \geqq H(\xi)\, , \quad H(\xi,\eta|\zeta) \geqq H(\xi|\zeta) \tag{25}$$

$$H(\xi|\eta,\zeta) \leqq H(\xi|\eta) \leqq H(\xi) \tag{26}$$

$$I(\xi_1,\xi_2 \wedge \eta) \geq I(\xi_1 \wedge \eta);\ I(\xi_1,\xi_2 \wedge \eta \mid \zeta) \geq$$

(27) $$\geq I(\xi_1 \wedge \eta \mid \zeta)$$

(28) $$I(\xi \wedge \eta) \leq H(\xi),\ I(\xi \wedge \eta \mid \zeta) \leq H(\xi \mid \zeta).$$

If ξ has at most r possible values then

(29) $$H(\xi) \leq \log_2 r.$$

If ξ has at most $r(y)$ possible values when $\eta = y$ then

(30) $$H(\xi \mid \eta) \leq E \log_2 r(\eta).$$

Proof. (25)-(28) are direct consequences of (19)-(24). (29) follows from (13) setting $\mathcal{P} = \mathcal{P}_\xi$, $\mathcal{Q} = \left\{\frac{1}{r}, \ldots, \frac{1}{r}\right\}$; on comparison of (6) and (8), (29) implies (30).

Remark. $I(\xi \wedge \eta \mid \zeta) \leq I(\xi \wedge \eta)$ is not valid; in general. E. g., if ξ and η are independent but not conditionally independent for a given ζ, then

$$I(\xi \wedge \eta) = 0 < I(\xi \wedge \eta \mid \zeta).$$

Theorem 3. (Substitutions in the information quantities).

For arbitrary functions $f(x)$, $f(y)$ or $f(x,y)$ defined on X, Y or $X \times Y$, respectively, the following inequalities hold

(31) $$H(f(\xi)) \leq H(\xi);\quad I(f(\xi) \wedge \eta) \leq I(\xi \wedge \eta)$$

$$H(\xi \mid f(\eta)) \geqq H(\xi \mid \eta) \tag{32}$$

$$H(f(\xi,\eta) \mid \eta) \leqq H(\xi \mid \eta). \tag{33}$$

If f is one-to-one, or $f(x,y)$ as a function of x is one-to-one for every fixed $y \in Y$, respectively, the equality signs are valid. In the second half of (31) and in (32) the equality holds also if ξ and η are conditionally independent for given $f(\xi)$ or $f(\eta)$, respectively.

Proof. In the one-to-one case, the validity of (31)-(33) with the equality sign is obvious from definition 2. In the general case, apply this observation for $\tilde{f}$ instead of f where $\tilde{f}(x) = (x, f(x))$, $\tilde{f}(y) = (y, f(y))$ or $\tilde{f}(x,y) = (x, f(x,y))$, respectively ; then theorem 2 gives rise to the desired inequalities. The last statements follow from (18) and the identities :

$$I(\xi \wedge \eta) = I(\xi, f(\xi) \wedge \eta) = I(f(\xi) \wedge \eta) + I(\xi \wedge \eta \ f(\xi))$$

$$H(\xi) = H(\xi, f(\xi)) \geqq H(f(\xi))$$

$$H(\xi \mid \eta) = H(\xi \mid \eta, f(\eta)) \leqq H(\xi \mid f(\eta))$$

$$H(\xi \mid \eta) = H(\xi, f(\xi,\eta) \mid \eta) \geqq H(f(\xi,\eta) \mid \eta)$$

respectively.

Theorem 4. (Convexity properties).

Consider the entropy and the mutual information as a function of the distribution of ξ, in the latter case keeping the conditional distributions $\mathcal{P}_{\eta|\xi=x} = \{p_x(y), y \in Y\}$ fixed :

(34) $$H(\mathcal{P}) = -\sum_{x\in X} p(x) \log_2 p(x)$$

(35) $$I(\mathcal{P}) = \sum_{x\in X, y\in Y} p(x)\, p_x(y) \log_2 \frac{p_x(y)}{q_{\mathcal{P}}(y)}\;;\; q_{\mathcal{P}}(y) = \sum_{x\in X} p(x) p_x(y).$$

Then $H(\mathcal{P})$ and $I(\mathcal{P})$ are concave functions of the PD $\mathcal{P} = \{p(x), x \in X\}$ i.e., if $\mathcal{P}_1 = \{p_1(x) = x \in X\}$, $\mathcal{P}_2 = \{p_2(x), x \in X\}$ and $\mathcal{P} = a\mathcal{P}_1 + (1-a)\mathcal{P}_2 = \{a p_1(x) + (1-a) p_2(x), x \in X\}$ where $0 < a < 1$ is arbitrary, then

(36) $$H(\mathcal{P}) \geqq a H(\mathcal{P}_1) + (1-a) H(\mathcal{P}_2), \quad I(\mathcal{P}) \geqq a I(\mathcal{P}_1) + (1-a) I(\mathcal{P}_2).$$

Proof. The function $f(t) = -t \log_2 t$ is concave hence so is $H(\mathcal{P})$ as well. Since the PD $\mathcal{Q}_{\mathcal{P}} = \{q_{\mathcal{P}}(y), y \in Y\}$ depends linearly on the PD $\mathcal{P}$, the concavity of $f(t) = -t \log_2 t$ also implies that

$$\sum_{x\in X} p(x)\, p_x(y) \log_2 \frac{p_x(y)}{q_{\mathcal{P}}(y)} =$$

$$= -q_{\mathcal{P}}(y) \log_2 q_{\mathcal{P}}(y) + \sum_{x\in X} p(x) p_x(y) \log_2 p_x(y)$$

is a concave function of $\mathcal{P}$, for every fixed $y \in Y$. Summation for all $y \in Y$ shows that $I(\mathcal{P})$ is concave, too.

Theorem 5. (Useful estimates with the I-divergence).

Let $\mathcal{P} = \{p(x), x \in X\}$ and $\mathcal{Q} = \{q(x), x \in X\}$ be two PD's on X. Then

$$\sum_{x\in X} |p(x) - q(x)| \leq \sqrt{\frac{2}{\log_2 e} I(\mathcal{P}\|\mathcal{Q})} \qquad (37)$$

$$\sum_{x\in X} p(x)\left|\log_2 \frac{p(x)}{q(x)}\right| \leq I(\mathcal{P}\|\mathcal{Q}) + \min\left(\frac{2\log_2 e}{e}, \sqrt{2\log_2 e \cdot I(\mathcal{P}\|\mathcal{Q})}\right). \qquad (38)$$

Proof. Let $A = \{x : p(x) \leq q(x)\}$, $B = \{x : p(x) > q(x)\}$; put $p(A) = p$, $q(A) = q$.
Then $p \leq q$, $p(B) = 1-p$; $q(B) = 1-q$,

$$\sum_{x\in X} |p(x) - q(x)| = 2(q-p), \qquad (39)$$

while from (11) and (12) it follows

$$I(\mathcal{P}\|\mathcal{Q}) \geq p \log_2 \frac{p}{q} + (1-p)\log_2 \frac{1-p}{1-q}. \qquad (40)$$

A simple calculation shows that

$$p \log_2 \frac{p}{q} + (1-p)\log_2 \frac{1-p}{1-q} - 2\log_2 e \cdot (p-q)^2 \geq 0 \qquad (41)$$
$$(0 \leq p \leq q \leq 1)$$

(for $p = q$ the equality holds and the derivative of the left hand side of (41) with respect to p is ≤ 0 if $0 < p \leq q < 1$).
The relations (39),(40), (41) prove (37).
From (11) and (12) it also follows

$$\sum_{x\in X} p(x)\left|\log_2 \frac{p(x)}{q(x)}\right| = I(\mathcal{P}\|\mathcal{Q}) - 2\sum_{x\in A} p(x)\log_2 \frac{p(x)}{q(x)} \leq$$
$$\leq I(\mathcal{P}\|\mathcal{Q}) - 2p\log_2 \frac{p}{q} = I(\mathcal{P}\|\mathcal{Q}) + 2p\log_2 \frac{q}{p}. \qquad (42)$$

Here

$$2p\log_2\frac{q}{p} = 2\log_2 e\cdot p\ln\frac{q}{p} \leqq 2\log_2 e\cdot p\ln\frac{1}{p} \leqq \frac{2\log_2 e}{e}$$

(since $f(t) = t\ln\frac{1}{t}$ takes on its maximum for $t = \frac{1}{e}$); further, more as $\ln\frac{q}{p} = \ln\left(1+\frac{q-p}{p}\right) \leqq \frac{q-p}{p}$, we also have (using (39)) $2p\log_2\frac{q}{p} \leqq 2\log_2 e\cdot(q-p) = \log_2 e\cdot\sum_{x\in X}|p(x)-q(x)|$. In view of these estimates, (42) and (37) imply (38).

1. Shannon's Theorem on Noiseless Channels.

A noiseless channel is a device which is able to transmit some signals $y_1, y_2, \dots, y_m$ (Channel signals) $(m \geq 2)$ from one place to another one, and the signals do not change during the transmission. The problem starts if we have an information source which emits some signals $x_1, x_2, \dots, x_n$ (information signals) $(n \geq 2)$ which are different from y_i's and we have to transmit x_i's by the channel. The only way to do this is to transform somehow the sequences of x_i's into sequences of y_i's which are transmissible by the channel. This transformations is called encoding. Obviously, it must be such a transformation, that we can uniquely restore the x_i's from the sequence of y_i's. We say in this case that the coding is uniquely decodable. No doubt, these notions require more precise definitions we shall give them later.

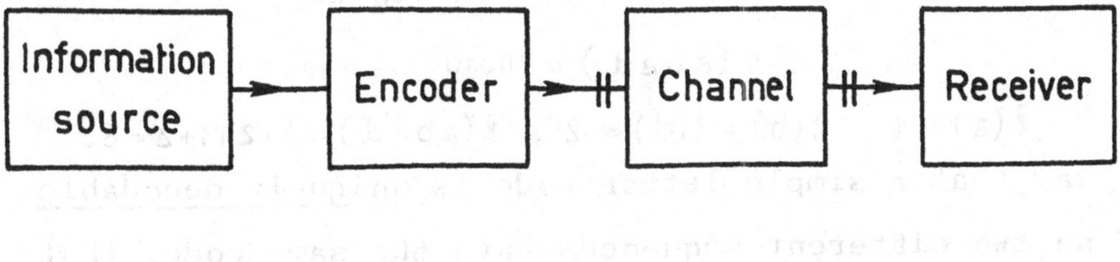

Our aim is to minimize the average (in a later defined sense) number of y_i 's needed to the transmission of one x_i in a uniquely decodable manner. First we try to characterize quantitatively the property of unique decodability.We start by definitions. Let us correspond with every x_i a finite sequence $y_{j1}, y_{j2}, \dots, y_{jh_i}$ which is marked by $c(x_i)$ and called the <u>code</u> of x_i . If we have a sequence $x_{i_1}, \dots, x_{i_k}$ its <u>code</u> is $c(x_{i_1} \dots x_{i_k}) = c(x_{i_1}) \dots c(x_{i_k})$. In other words we form the code of a sequence by simple juxtaposition of the codes of the elements of the sequence. The transformation defined here is called <u>simple letter code</u>.

$$\ell(u) = \ell(x_{i_1} \dots x_{i_k}) = \sum_{j=1} \ell(x_{i_j}) = \sum_{j=1} h_{i_j}$$

denotes the <u>code length</u>, that is, the number of y_i's contained in the code of $u = (x_{i_1}, \dots, x_{i_k})$.

<u>Example 1</u>. Let $X = \{x_1, x_2, x_3\} = \{a, b, d\}$ and $Y = \{y_1, y_2\} = \{0,1\}$. Fix the codes of the letters a, b, d in the following manner :

$$c(a) = 1, \quad c(b) = 00 \quad c(d) = 01.$$

Then, for instance the code of $abad$ is

$$c(abad) = 100101.$$

$$\ell(a) = 1, \quad \ell(b) = \ell(d) = 2, \quad \ell(abad) = 1+2+1+2 = 6.$$

We say that a simple letter code is <u>uniquely decodable</u> if no two different sequences have the same code, that is, $c(x_{i_1} \ldots x_{i_r}) = c(x_{j_1} \ldots x_{j_s})$ may hold only if $r = s$ and $i_k = j_k$ $(1 \le k \le r)$. If $y_{i_1} \ldots y_{i_p} = u$ is a sequence then $y_{i_1} \ldots y_{i_q}$ $(0 \le q \le p)$ is <u>segment</u> of u. The void sequence is a segment of everything. Now, we say that a simple letter code is <u>prefix</u> if no $c(x_i)$ is a segment of an other $c(x_j)$ $(i \ne j)$.

<u>Lemma 1. A prefix simple letter code is uniquely decodable.</u>

The formal <u>proof</u> of this triviality is the following. Let $c(x_{i_1} \ldots x_{i_r}) = c(x_{j_1} \ldots x_{j_s})$ $(r \le s)$. By definition, $c(x_{i_1})$ is a segment of $c(x_{j_1})$ or conversely, $c(x_{j_1})$ is a segment of $c(x_{i_1})$. Hence, by the prefix property $i_1 = j_1$, and $c(x_{i_2} \ldots x_{i_p}) = c(x_{j_2} \ldots x_{j_s})$. Repeating this procedure we get $i_2 = j_2, \ldots, i_r = j_r$. If $r = s$, we are ready. If $r < s$, we get $c(x_{j_{r+1}}), \ldots, c(x_{j_s})$ are void sequences which are segments of everything, in contradiction with the prefix property. The proof is completed. However there are uniquely decodable simple letter codes which are not prefix.

<u>Example 2</u>. Put

$$c(a) = 1 \qquad c(b) = 00 \qquad c(d) = 10.$$

The code is not prefix, for $c(a)$ is a segment of $c(d)$. But we may uniquely determine the first information signal from the code sequence. If the first code letter is 0 ,then, the first information letter is b. If the first code letter is 1 ,then the first information letter is a or b depending on the parity of the number of 0's standing after the first 1 . When we determined the first information signals, we may continue the procedure similarly. For instance, if the code sequence is 100110 ,then the information sequence is uniquely abad .

Now we are able to formulate the promised quantitative charaterization of the uniquely decodability.

Theorem 1. From the information signals $x_1, \ldots, x_n$ into the channel (or code) signals $y_1, \ldots, y_m$ there exists a uniquely decodable simple letter code with fixed code lengths $c(x_1) = h_1, \ldots, c(x_n) = h_n$ if and only if

$$\sum_{i=1}^{n} \frac{1}{m^{h_i}} \leq 1 . \tag{1.1}$$

It is the so called Kraft inequality, but several authors use Mac-Millan's, Fano's and Szilárd's name with this connection.

Proof. We shall proof that if there exists a uniquely decodable code with code lengths $h_1, \ldots, h_n$, then (1.1) holds and conversely, if (1.1) holds then there exists a prefix simple letter code with code lengths $h_1, \ldots, h_n$. The first part is the more interesting one. Several proofs are known. Here we give the proof of Karush [1]. First we prove the equality

$$\sum_{K=1}^{hN} w(N,k)m^{-K} = (m^{-h_1} + \ldots + m^{-h_n})^N, \qquad (1.2)$$

where $w(N,k)$ denotes the number of sequences $x_{i_1},\ldots, x_{i_N}$ for which the code length is $\ell(x_{i_1}\ldots x_{i_N}) = k$, and $h = \max(h_1,\ldots,h_n)$. We shall use the following formula

$$w(N+1,k) = \sum_{i=1} w(N,K-h_i) \qquad (1.3)$$

which is a simple consequence of the fact, that the code of $N+1$ x_i's starts with a $c(x_i)$ and the remaining code sequence is a code of N x_i's. Now we can prove (1.2) by induction over N. For $N=1$ the statement is trivial. If $N+1>1$, then we get

$$\left(\sum_{i=1}^{n} m^{-h_i}\right)^{N+1} = \left(\sum_{K=1}^{hN} w(N,K)m^{-K}\right)\left(\sum_{i=1}^{n} m^{-h_i}\right) =$$

$$= \sum_{K=1}^{hN}\sum_{i=1}^{n} w(N,K)m^{-k-h_i} = \sum_{i=1}^{n}\sum_{K=1}^{hN} w(N,K)m^{-K-h_i} =$$

$$= \sum_{k=1}^{h(N+1)} m^{-K}\sum_{i=1}^{n} w(N,K-h_i) = \sum_{K=1}^{h(N+1)} m^{-k} w(N+1,K)$$

by the induction hypothesis and (1.3). Since the number of code sequences of length k is at least m^k, so $w(N,K) \leq m^K$ that is, $w(N,K)m^{-K} \leq 1$. Summing this, we obtain

$$\sum_{K=1}^{hN} w(N,K)m^{-K} \leq hN \tag{1.4}$$

and (1.4) results

$$\left(\sum_{i=1}^{n} m^{-h_i}\right)^N \leq hN. \tag{1.5}$$

If $\sum_{i=1}^{n} m^{-h_i} > 1$, then the left hand side tends to the infinity faster then the right hand side, when $N \to \infty$ what contradicts (1.5). So, we get the desired Kraft inequality.

Assume now that we have the fixed numbers $h_1 \leq h_2 \leq \ldots \leq h_n = h$ satisfying the Kraft inequality. Choose for $c(x_1)$ an arbitrary sequence of length h_1. Denote by A_1 the set of all sequences of length h which contain $c(x_1)$ as a segment. In other words A_1 consists of the sequences obtained from $c(x_1)$ by writing $h-h_1$ arbitrary y_i's to the end. $|A_1|$ denotes the number of the elements of A_1. It is obvious, that $|A_1| = m^{h-h_1}$. Rewriting (1.1) we get

$$\sum_{i=1}^{n} m^{h-h_i} \leq m^h. \tag{1.6}$$

If $n > 1$ (if $n=1$ the statement is trivial) then $|A_1| < m^h$, thus we may choose a sequence of length h which is not in A_1. The first h_2 letters of this sequence will form $c(x_2)$. It is easy to see, that $c(x_1)$ can not be a segment of $c(x_2)$ (and conversely). We can define A_2 similarly, and A_1 and A_2 are disjoint. The number $|A_2|$ is m^{h-h_2}, thus we get $|A_1|+|A_2|<m^h$, too (if $n>2$). We can again choose a sequence of length h which is not an element of A_1 or A_2. The first h_3 letters of this sequence form $c(x_3)$. We can continue this procedure up to n. The proof is completed.

In order to minimize the "average" code length we have to define the meaning of the word "average". The information source $\mathfrak{X}$ emits independent random signals; with probability p_i the signal x_i. Now, we can define the average code length as

$$L = \sum p_i h_i ,$$

where $h_i = \ell(x_i)$.

It would be very nice to have a measure of information which proves the material-like property of information. As everybody knows, the Shannon entropy

$$H(\mathfrak{X}) = -\sum_{i=1}^{n} p_i \log p_i \qquad (1.7)$$

satisfies this requirement ; it has some properties of this type. The first one is the following. The maximum of the entropy (1.7) for m different signals is $\log m$, so one code signal (y_i) may contain at most $\log m$ of information. One information signal contains $H(\mathfrak{X})$ information, thus the average number of code signals necessary to transmit one information signal is at least $\frac{H(\mathfrak{X})}{\log m}$, that is :

(1.8) $$\min L \sim \frac{H(\mathfrak{X})}{\log m} .$$

We shall investigate (1.8) in these pages.

Theorem 2. If $\mathfrak{X}$ is an information source which emits independently the letters $x_1, \ldots, x_n$ with probabilities $p_1, \ldots, p_n$, further there is a uniquely decodable simple letter code $c(x_1), \ldots, c(x_n)$ with code lengths $h_1, \ldots, h_n$ and average code length $L = \sum_{i=1}^{n} p_i h_i$ then

(1.9) $$L \geq \frac{H(\mathfrak{X})}{\log m} .$$

Proof. From the introduction we know

by (11) and (13) $\sum p_i \log \frac{p_i}{q_i} \geq 0$

or

(1.10) $$- \sum p_i \log p_i \leq - \sum p_i \log q_i ,$$

where $q_i \geq 0$ $(1 \leq i \leq n)$, $\sum_{i=1}^{n} q_i = 1$. Choosing

$$q_i = \frac{m^{-h_i}}{\sum_{k=1}^{n} m^{-h_k}}$$

we get from (1.10)

$$H(\mathfrak{X}) \leq -\sum_{i=1}^{n} p_i \log m^{-h_i} + \sum_{i=1}^{n} p_i \log\left(\sum_{i=1}^{n} m^{-h_i}\right). \qquad (1.11)$$

For the code is uniquely decodable we may use the Kraft inequality

$$\log\left(\sum_{k=1}^{n} m^{-h_k}\right) \leq 0.$$

Substituting it into (1.11) we obtain

$$H(\mathfrak{X}) \leq L \cdot \log m,$$

what is equivalent to (1.9).

Theorem 2 answers only one side of our problem (1.8). However, we do not know whether we can reach the bound $\frac{H(\mathfrak{X})}{\log m}$. The following theorem is the first step in this direction.

Theorem 3. If $\mathfrak{X}$ is an information source which emits independently the letters $x_1, \ldots, x_n$ with probabilities $p_1, \ldots, p_n$, then there exists a uniquely decodable (prefix) simple letter code from the code letters $y_1, \ldots, y_m$ with average code length L satisfying the inequality

$$\frac{H(\mathfrak{X})}{\log m} \leq L \leq \frac{H(\mathfrak{X})}{\log m} + 1. \qquad (1.12)$$

Proof. Let $\{a\}$ denote the least integer $\geq a$.

Let us choose $h_i = \left\{ -\frac{\log p_i}{\log m} \right\}$. As

$$\sum_{i=1}^{n} m^{-\left\{-\frac{\log p_i}{\log m}\right\}} \leq \sum_{i=1}^{n} m^{-\left(-\frac{\log p_i}{\log m}\right)} = \sum_{i=1}^{n} p_i = 1,$$

they satisfy the Kraft inequality. Thus, by Theorem 1 there exists a uniquely decodable prefix simple letter code with code lengths $h_1, \dots, h_n$. The average code length satisfies the following inequality :

$$L = \sum_{i=1}^{n} p_i \left\{ -\frac{\log p_i}{\log m} \right\} \leq \sum_{i=1}^{n} p_i \left(-\frac{\log p_i}{\log m} + 1 \right) = \frac{H(\mathfrak{X})}{\log m} + 1.$$

This is the right hand side of (1.12). The left hand side is the consequence of Theorem 2. The proof is completed.

Theorem 3 shows that we can approximate the lower bound $\frac{H(\mathfrak{X})}{\log m}$ rather well with simple letter codes. However, we can not reach it, as the following example shows.

<u>Example 3</u>. Put $n = 2$ $p_1 = \frac{1}{3}$, $p_2 = \frac{2}{3}$. $H(\mathfrak{X}) < 1$ and $\frac{H(\mathfrak{X})}{\log m} < 1$ are obvious. Similarly, $L \geq 1$ is trivial, thus $\frac{H(\mathfrak{X})}{\log m} < L$, indeed.

We must use other codes to reach the lower bound. Let us consider the sequence $x_{i_1}, \dots, x_{i_N}$ of length N and let us define its code $c(x_{i_1} \dots x_{i_N})$ as a fixed sequence of y_i's. The code of a sequence $x_{i_1} \dots x_{i_K}$ is

$$c(x_{i_1} \dots x_{i_K}) = c(x_{i_1} \dots x_{i_N})\, c(x_{i_{N+1}} \dots x_{i_{2N}}) \dots c(x_{i_{(q-1)N+1}} \dots x_{i_{qN}}),$$

where $K = qN+r$, $0 \leq r < N$. This code called block code, or N block code. We have to redefine the notion of average code length (of one information signal). The average code length of N information signals is the same as earlier :

$$L_N = \sum_{\substack{1 \leq i_1 \leq n \\ \vdots \\ 1 \leq i_N \leq n}} p_{i_1} \cdots p_{i_N} \, \ell(x_{i_1} \cdots x_{i_N}).$$

The average code length (of one information signal) is $L = \frac{L_N}{N}$. We may approximate the lower bound (1.8) with block codes better, then with simple letter codes.

Theorem 4. Let $\mathcal{X}$ be an information source which emits independently the letters $x_1, \dots, x_n$ with probabilities $p_1, \dots, p_n$. If we have a block code made from the code letters $y_1, \dots, y_m$ and the average code length is L, then

$$\frac{H(\mathcal{X})}{\log m} \leq L . \tag{1.13}$$

On the other hand, if an $\varepsilon > 0$ is given, then there exists an N-block code satisfying

$$L \leq \frac{H(\mathcal{X})}{\log m} + \varepsilon \tag{1.14}$$

(for $N \geq \frac{1}{\varepsilon}$).

Proof. We may consider the sequences $x_{i_1} \dots x_{i_N}$ as new information signals. In this case the block code becomes simple letter code. Denoting by $\mathfrak{X}^N$ the new source, we may use Theorems 2 and 3 : for every code

$$\frac{H(\mathfrak{X}^N)}{\log m} \leq L_N \tag{1.15}$$

holds and there is a code with the property

$$L_N \leq \frac{H(\mathfrak{X}^N)}{\log m} + 1 .$$

But $H(\mathfrak{X}^N) = N H(\mathfrak{X})$, thus dividing by N we get (1.13) from (1.15) and (1.14) from (1.16). $H(\mathfrak{X}^N) = N H(\mathfrak{X})$ follows from the fact that the information signals are independent and for independent random variables ξ_1, ξ_2 $H(\xi_1,\xi_2) = H(\xi_1) + H(\xi_2)$ rather than (2.19). The proof is completed.

However, we cannot reach the lower bound in general even with block codes. If $n = 2$, p_1 and p_2 are rational numbers, then L_N and L are rational numbers, but $H(\mathfrak{X})$ can be irrational. This is the case if $p_1 = \frac{1}{3}$ and $p_2 = \frac{2}{3}$.

On the basic properties of noiseless channels see e.g. [9].

2. The Principle of Conservation of Entropy.

Let us go back to (1.8). If it is (approximately) true for the best possible encoding, then it must be true for arbitrary uniquely decodable coding. After the encoding we get a new random sequence of signals ; we may consider it as a new information source. Let us denote by $\mathcal{Y}$. Further, $H(\mathcal{Y})$ denotes the information contant of one signal of $\mathcal{Y}$ (for a moment heuristically). If the information is material-like, then the information contant of one information signal must be distributed on the L code signals which transmit it :

$$\frac{H(\mathcal{X})}{L} = H(\mathcal{Y}). \qquad (2.17)$$

It is called the principle of conservation of entropy [2] and we shall now formulate precisely and prove it.

The code sequence consists of random variables, however, they are not independent. Thus we have to define the entropy of an arbitrary discrete stochastic process. Let $\xi(1),\dots,\xi(k),\dots$ be a stochastic process* with state space $x=\{x_1,\dots,x_n\}$. This is the more general definition of the information source $\mathcal{X}$. We use the following notation $\xi(\ell,k)=(\xi(\ell),\dots,\xi(k))$ $(\ell\leq k)$, thus $P(\xi(1)=x_{i_1},\dots,\xi(k)=x_{i_k}) = P(\xi(1,k)=u)$, where $u=(x_{i_1},\dots,x_{i_k})$. The entropy of the first N signal is

$$H(\xi(1,N)) = -\sum_{u\in X^N} P(\xi(1,N)=u)\log P(\xi(1,N)=u).$$

*) That is, $\xi(1),\dots,\xi(k),\dots$ are random variables with possible values $x_1,\dots,x_n$.

Now, we are able to define the <u>entropy per simbol</u> of an information source $\mathfrak{X}$:

$$(2.18) \qquad H(\mathfrak{X}) = \lim_{N\to\infty} \frac{H(\xi(1,N))}{N},$$

if this limit exists. When does there exist such a limit ? We give here a sufficient condition. An information source is <u>stationary</u>, if

$$P(\xi(\ell,k)=u) = P(\xi(1,k-\ell+1)=u)$$

for every $1 \leq \ell \leq k$ integer.

<u>Theorem 5. If</u> $\mathfrak{X}$ <u>is a stationary information source, then</u> $H(\mathfrak{X})$ (<u>that is, the limit) (2.18</u>) <u>exists</u>.

To the proof we need a lemma.

<u>Lemma 2</u>. <u>If</u> $H(j) \geq 0$ <u>is a monotone increasing real function defined on the natural numbers and it satisfies the inequality</u>

$$(2.19) \qquad H(j+k) \leq H(j) + H(k),$$

<u>then the limit</u>

$$\lim_{k\to\infty} \frac{H(k)}{k}$$

<u>exists</u>.

<u>Proof</u>. Put

$$\underline{\lim}_{k\to\infty} \frac{H(k)}{k} = a.$$

We have to verify that for arbitrary $\varepsilon > 0$ there is a threshold $N(\varepsilon)$ such that for $N > N(\varepsilon)$

$$\frac{H(N)}{N} \leq a+\varepsilon \qquad (2.20)$$

holds. Choose a k_0 satisfying

$$\frac{H(k_0)}{k_0} \leq a + \frac{\varepsilon}{2} \ .$$

Let $N = qk_0 + r$, $0 \leq r < k_0$ be a natural number. Then, by (2.19) and the monotonicity we have

$$\frac{H(N)}{N} \leq \frac{H(qk_0)+H(r)}{qk_0+r} \leq \frac{qH(k_0)+H(k_0)}{qk_0} = \frac{q+1}{q}\,\frac{H(k_0)}{k_0} . \qquad (2.21)$$

For large enough q

$$\frac{q+1}{q}\,\frac{H(k_0)}{k_0} \leq a+\varepsilon$$

holds, what together with (2.21) proves (2.20) and the lemma. (See [9]).

<u>Proof of theorem 5.</u> Put $H(j) = H(\xi(1,j))$ and use the lemma. We have to verify only the monotonicity and the property (2.19) for $H(j)$:

$$H(\xi(1,i)) \leq H(\xi(1,j)) = H(\xi(1,i),\xi(i+1,j)) \quad (i \leq j)$$

and

$$H(\xi(1,j+k)) = H(\xi(1,j),\xi(j+1,j+k)) \leq H(\xi(1,j)) +$$
$$+ H(\xi(j+1,j+k)) = H(\xi(1,j)) + H(\xi(1,k)) .$$

They follows from

$$H(\xi_1) \leq H(\xi_1, \xi_2) \leq H(\xi_1) + H(\xi_2)$$

which is true for arbitrary discrete random variables ξ_1, ξ_2 by (2.19) and (2.26).

If $\mathfrak{X}$ is an independent or stationary, then $\mathfrak{Y}$ is not necessarily independent or stationary. Thus, it seems, the best is to consider general information sources. However, we did not define the average code length for general source yet. We do here. If a code is given, $L_N = \ell(\xi(1) \ldots \xi(N))$ is the code length of the first N information signal. It is a random variable. We say that L is the average code length if

$$\frac{L_N}{N} \Longrightarrow L,$$

where $\Longrightarrow$ means the convergence in probability, that is, if

$$\lim_{N \to \infty} P\left(\left|\frac{L_N}{N} - L\right| > \varepsilon\right) = 0$$

for every $\varepsilon > 0$. We do not want to investigate here when does this L exist, because it is not an information theoretical question. We just mention if $\mathfrak{X}$ is a stationary and ergodic process, then L exists.

If an information source $\mathfrak{X}$ and a code c are given, then they determine a random sequence of code signals. We denote these signals by $\eta(1),\dots,\eta(k),\dots$ and the stochastic process by $\mathcal{Y}$. The information contant of one signal is again determined by

$$H(\mathcal{Y}) = \lim_{N\to\infty} \frac{H(\eta(1,N))}{N},$$

if this limit exists.

After these preparations we may start to prove (2.17).

<u>Theorem 6. Let</u> $\mathcal{X}$ <u>be an information source and let</u> c <u>be a uniquely decodable simple letter code from</u> $\mathcal{X}$ <u>into sequences of signals</u> $y_1,\dots,y_m$. <u>Assume the entropy</u> $H(\mathcal{X})$ <u>and the average code length</u> $L>0$ <u>exist. Then</u> $H(\mathcal{Y})$ <u>also exists and</u>

$$H(\mathcal{Y}) = \frac{H(\mathcal{X})}{L}. \tag{2.22}$$

In order to prove this theorem we need a new notation and a lemma. Let $\nu(N)$ denote the random variable defined by

$$\ell(x_{i_1},\dots,x_{i_{\nu(N)}}) \le N < \ell(x_{i_1},\dots,x_{i_{\nu(N)+1}}).$$

<u>Lemma 3. Let</u> $\mathcal{X}$ <u>be an information source and let</u> c <u>be a simple letter code (not necessarily uniquely decodable) but</u> $\ell(x_i)>0$ $(1\le i\le n)$ <u>from</u> $\mathcal{X}$ <u>into sequences of signals</u> $y_1,\dots,y_m$. <u>If</u> $L>0$ <u>exists, then</u>

$$\lim_{N\to\infty}\left(\frac{H(\xi(1,k))}{L\cdot K} - \frac{H(\xi(1,\nu(N)))}{N}\right) = 0, \tag{2.23}$$

where $k = \left[\frac{N}{L}\right]$.

Proof. We start the proof with the equality

$$H(\xi(1,k)) + H(\xi(1,\nu(N)) \mid \xi(1,k)) =$$

$$= H(\xi(1,\nu(N))) + H(\xi(1,k) \mid \xi(1,\nu(N))),$$

what follows from (2.19) and its converse $H(\xi,\eta) = H(\eta|\xi) + H(\xi)$. It results

$$\frac{H(\xi(1,k))}{N} - \frac{H(\xi(1,\nu(N)))}{N} =$$

$$= \frac{H(\xi(1,k)|\xi(1,\nu(N)))}{N} - \frac{H(\xi(1,\nu(N))|\xi(1,k))}{N},$$

thus, to (2.23) it is sufficient to prove

(2.24) $$H(\xi(1,\nu(N))|\xi(1,k)) = o(N)$$

and

(2.25) $$H(\xi(1,k)|\xi(1,\nu(N))) = o(N),$$

where $k = \left[\frac{N}{L}\right]$. However,

(2.26) $$H(\xi(1,\nu(N))|\xi(1,k)) = \sum_{u \in X^N} P(\xi(1,k) = u)\, H(\xi(1,\nu(N))|u),$$

where $H(\xi(1,\nu(N))|u)$ denotes the entropy

$$H(\xi(1,\nu(N))|u) =$$

$$= \sum_{u' \in X(N)} P(\xi(1,\nu(N)) = u' | \xi(1,k) = u) \cdot \log P(\xi(1,\nu(N)) = u' | \xi(1,k) = u)$$

and $X(N)$ denotes the state space of $\xi(1,\nu(N))$.

If $\ell(u) \geq N$ then $P(\xi(1,\nu(N)) = u' | \xi(1,k) = u) \neq 0$ can hold only if u' is a segment of u (as u cannot be a proper segment of u'), however there can be just one segment of u in $X(N)$ because of definition of $X(N)$. Thus, $P(\xi(1,\nu(N)) = u' | \xi(1,k) = u) = 1$ for a certain u', and

$$H(\xi(1,\nu(N))| u) = 0. \tag{2.27}$$

Assume now $\ell(u) < N$, $P(\xi(1,\nu(N)) = u' | \xi(1,k) = u) \neq 0$ if and only if u' is a segment of u or u is a segment of u'. If u' is a proper segment of u, then $\ell(u') < \ell(u) < N$ contradicts the definition of $\nu(N)$. Thus, it remains the case u is a segment of u'. The number of such sequences u' is at most $n^{N-\ell(u)}$, since lengthening u by $N - \ell(u)$ signals the code is lengthened at least by $N - \ell(u)$ code signals (the code length of one signal > 0); the code length of a lengthened sequence $\geq N$, thus all the sequences are segments of these sequences. Hence we get the estimation

$$H(\xi(1,\nu(N))| u) \leq \log n^{N-\ell(u)} = (N - \ell(u)) \log n . \tag{2.28}$$

Applying (2.26), (2.27) and (2.28) we have

$$H(\xi(1,\nu(N))|\xi(1,k)) = \sum_{\substack{u\in X^k\\ \ell(u)<N}} P(\xi(1,k)=u)H(\xi(1,\nu(N))|u) \le$$

$$\le \sum_{\substack{u\in X^k\\ \ell(u)<N}} P(\xi(1,k)=u)(N-\ell(u))\log n = \sum_{\substack{u\in X^k\\ N(1-\varepsilon)\le \ell(u)<N}} + \sum_{\substack{u\in X^k\\ \ell(u)<(1-\varepsilon)N}} \le$$

$$\le N\varepsilon \log n + P(u\in X^k, \ell(u)<(1-\varepsilon)N)\, N\log n .$$

Hence

$$(2.29)\qquad \frac{H(\xi(1,\nu(N))|\xi(1,k))}{N} \le \varepsilon \log n + P(u\in X^k, \ell(u)<(1-\varepsilon)N)\log n$$

follows. Since $\frac{\ell(u)}{k}$ converges in probability to L, $P(u\in X_1^k \mid \frac{\ell(u)}{k} - L| > \varepsilon)$ converges to zero if $k\to\infty$. It follows that on the right side of (2.29)

$$P(u\in X^k, \ell(u)<(1-\varepsilon)N) = P(u\in X^k, \frac{\ell(u)}{k} - L < (1-\varepsilon)\frac{N}{k} - L) \le$$

$$\le P(u\in X^k, \frac{\ell(u)}{k} - L < -L\cdot\frac{\varepsilon}{2})$$

tends to zero for $N\to\infty$ because of $k=\left[\frac{N}{L}\right]$. Thus, if N is sufficiently large,

$$P(u\in X^k, \ell(u)<(1-\varepsilon)N) < \varepsilon ,$$

that is,

$$\frac{H(\xi(1,\nu(N))|\xi(1,k))}{N} \le 2\varepsilon \log n ,$$

consequently

$$\lim_{N\to\infty} \frac{H(\xi(1,\nu(N))|\xi(1,k))}{N} = 0 \qquad \left(k=\left[\frac{N}{L}\right]\right),$$

and (2.24) is proved. We shall prove (2.25) in a similar way. Obviously,

$$H(\xi(1,k)|\xi(1,\nu(N))) = \sum_{u'\in X(N)} P(\xi(1,\nu(N))=u')\,H(\xi(1,k)|u'),$$

where $H(\xi(1,k)|u')=0$ holds in the case $\nu(N)\geq k$ $(u'=(x_{i_1},\dots,x_{i_{\nu(N)}}))$. Further, if $\nu(N)<k$ we have $n^{k-\nu(N)}$ different sequences u with $P(\xi(1,k)=u|\xi(1,\nu(N))=u')\neq 0$, that is, $H(\xi(1,k)|u')\leq \log n^{k-\nu(N)} = (k-\nu(N))\log n$. Finally, as in the above part we obtain

$$H(\xi(1,k)|\xi(1,\nu(N)))\leq \sum_{\substack{u'\in X(N)\\ k(1-\varepsilon)\leq\nu(N)<k}} (P\ \xi(1,\nu(N))=u')(k-\nu(N))\log n +$$

$$+\sum_{\substack{u'\in X(N)\\ \nu(N)<(1-\varepsilon)k}} P(\xi(1,\nu(N))=u')\,(k-\nu(N))\log n \leq$$

$$\leq \varepsilon k\log n + k\log n\ P(u'\in X(N),\nu(N)<(1-\varepsilon)k)\,.$$

Here on the right hand side

$$P(u'\in X(N),\nu(N)<(1-\varepsilon)k)\leq P(u\in X^{[(1-\varepsilon)k]},\ \ell(u)\geq N-h)$$

is true, where $h=\max(\ell(x_1),\dots,\ell(x_n))$. Further, by simple calculations we get

$$P(u\in X^M,\ \ell(u)\geq N-h) = P\left(u\in X^M,\ \frac{\ell(u)}{M}-L\geq \frac{N-h}{(1-\varepsilon)k}-L\right)\leq$$

$$\leq P\left(u\in X^M,\ \frac{\ell(u)}{M}-L\geq \frac{L}{1-\varepsilon}-L-\frac{h}{(1-\varepsilon)k}\right)\leq$$

$$\leq P\left(u\in X^M,\ \frac{\ell(u)}{M}-L\geq\frac{\varepsilon}{2}L\right)$$

which converges to zero if $M=[(1-\varepsilon)k]\to\infty$. Thus $H(\xi(1,k)|\xi(1,\nu(N)))= o(N)$, indeed, what proves (2.25) and the lemma.

<u>Proof of theorem 6</u>. If $H(\mathcal{X})$ exists, in Lemma 3 $\frac{H(\xi(1,k))}{L\cdot k}$ tends to $\frac{H(\mathcal{X})}{L}$ and thus $\frac{H(\xi(1,\nu(N)))}{N}$ does, too. Thus, it is sufficient to show that

$$\lim_{N\to\infty} \frac{H(\xi(1,\nu(N)))}{N} = \lim_{N\to\infty} \frac{H(\eta(1,N))}{N} .$$

We use the same idea as in the proof of lemma 3 :

$$H(\eta(1,N)) + H(\xi(1,\nu(N)) \mid \eta(1,N)) =$$

(2.30) $$= H(\xi(1,\nu(N)) + H(\eta(1,N) \mid \xi(1,\nu(N))).$$

Obviously, $P(\eta(1,N)=v, \xi(1,\nu(N))=u')\neq 0$ only if $c(u')$ is a segment of v. Thus, for a given v there are at most h $c(u')$ satisfying $P(\eta(1,N)=v, \xi(1,\nu(N))=u')=0$, because $N-h<\ell(u')\leq N$, that is by uniquely decodability there are at most h u' with $P(\eta(1,N)=v, \xi(1,\nu(N))=u')=1$. Applying this result we obtain

$$H(\xi(1,\nu(N)) \mid \eta(1,N)) =$$

(2.31) $$= \sum_{v\in Y^N} P(\eta(1,N)=v) H(\xi(1,\nu(N)) \mid v) \leq \log h$$

from $H(\xi(1,\nu(N))|v)\leq \log h$, which tends to 0 divided by N. On the other hand if we fix u' then, because of $N-h < \ell(u') \leq N$, there are at most m^h sequences $v\in Y^N$ for which $c(u')$ is a segment of v. Hence we have

$$H(\eta(1,N) \mid u') \leq \log m^h \quad \text{and} \quad H(\eta(1,N) \mid \xi(1,\nu(N))) =$$

$$= \sum P(\xi(1,\nu(N)) = u') H(\eta(1,N)|u') \leq \sum P(\xi(1,\nu(N))=u') \log m^h = h \log m$$

which tends to zero divided by N if $N\to\infty$. The proof is finished by (2.30) and (2.31).

It is clear that $H(\eta(1,N)) \leq \log m^N$ and hence

$$H(\mathcal{Y}) = \lim_{N\to\infty} \frac{H(\eta(1,N))}{N} \leq \log m .$$

Using (2.22) we get

$$L \geq \frac{H(\mathcal{X})}{\log m}$$

if the assumptions of Theorem 6 hold. Obviously, it is a generalization of Theorem 2 for more general information sources. Our important aim is to prove this for more general codes and channels.

The statement (2.22), dating back to Shannon [11], has often been regarded as "obvious". It is proved rigorously in [2]. A special case is proved in [10].

If the coding is not necessarily uniquely decodable, we cannot prove the existence of $H(\mathcal{Y})$. In this case let us put

$$\bar{H}(\mathcal{Y}) = \overline{\lim_{N \to \infty}} \frac{H(\mathcal{Y}^N)}{N} .$$

In this case from the above we get only $H(\eta(1,N)) \leq$ $\leq H(\xi(1,\nu(N)))$, that is, the following theorem holds.

<u>Theorem 7. Let</u> $\mathcal{X}$ <u>be an information source and let</u> c <u>be a simple letter code from</u> $\mathcal{X}$ <u>into sequences of signals</u> $y_1, \dots, y_m$. <u>If the entropy</u> $H(\mathcal{X})$ <u>and the average code length</u> $L > 0$ <u>exists, then</u>

$$\bar{H}(\mathcal{Y}) \leq \frac{H(\mathcal{X})}{L} .$$

3. Entropy with Respect to a Cost Scale.

Let us inspect Lemma 3 once more. We may consider $\xi(1, \nu(N))$ in the following way. The time for the information signals is measured according to their code length, and we ask what happens according to this time-scale up to N. $H(\xi(1,\nu(N)))$ is the entropy of this event and

$$\lim_{N \to \infty} \frac{H(\xi(1, \nu(N)))}{N}$$

is the entropy (per unit time) of the source according to this time-scale. There are many examples of this type. Assume we know the entropy of the written text. The durations of the different letters in the spoken text are different. Thus, if we want to know the entropy (per unit time) of the spoken text, it is suitable to measure it with respect to the time-scale of the durations. If the text is encoded by Morse-alphabet we may also consider the entropy per unit time, since the 3 Morse-signals have different lengths. There are examples when some other costs are important rather then durations. Therefore we use the word cost-scale for this notion. On the basis of these examples we could define the cost-scales as non-negative numbers

$s_1, \ldots, s_n$ corresponded with $x_1, \ldots, x_n$. However we need a more general definition as the following example shows. Let us consider the cost-scale generated by the code lengths for an N-block code. It is clear by definition of block codes that the code $c(x_{i_1}, \ldots, x_{i_{N-1}})$ of the first $N-1$ information signals is the void sequence u_0, thus the "duration" of the first N signals with respect to this cost-scale is zero. The "duration" of the first N signals is the length $\ell(x_{i_1}, \ldots, x_{i_N})$, and so on. Here the "durations" depend not only on the last signal.

A cost-scale σ on the information source $\mathfrak{X}$ is a real function $s(u)$ defined on the finite sequences $u = x_{i_1} \ldots x_{i_k}$ such that s (void sequence) $= 0$ and $s(u) \leq s(v)$ if u is a segment of v and $\lim_{k \to \infty} s(u_k) = \infty$, where u_k is a proper segment of u_{k+1}. The particular case when

$$s(u) = \sum_{j=1}^{k} s_{i_j}$$

is called memoryless cost-scale. (For more general cost-scales see [3]). The random variable $\nu(t)$ is defined by

$$s(\xi(1, \nu(t))) \leq t < s(\xi(1, \nu(t) + 1)).$$

The entropy per unit cost with respect to the cost-scale σ is

$$H(\mathfrak{X} \| \sigma) = \lim_{t \to \infty} \frac{H(\xi(1, \nu(t)))}{t}$$

if this limit exists.

A particular cost-scale is the counting-scale γ for which

$$\gamma(u) = k \quad \text{where} \quad u = x_{i_1} \dots x_{i_k} .$$

Lemma 3 gives the relation between the entropy with respect to the counting-scale and the entropy with respect to the cost-scale defined by a simple letter code. Now we shall examine a more general problem. What is the relation between the entropies with respect to two different cost-scales. If σ is a cost-scale, and r is a real number then $r\,\sigma$ means the cost-scale de-. fined by

$$s^r(u) = r\, s(u) \qquad \left(\nu^r(t) = \nu\left(\frac{t}{r}\right)\right).$$

Theorem 8. Let σ_1 and σ_2 be two cost-scales on the information source $\mathfrak{X}$. If one of the entropies $H(\mathfrak{X} \| \sigma_1)$ and $H(\mathfrak{X} \| \sigma_2)$ exists further

(3.32) $$\lim_{t \to \infty} \frac{E\,|\nu_1(t) - \nu_2^r(t)|}{t} = 0 \quad \text{holds for some } r > 0,$$

then

(3.33) $$r\,H(\mathfrak{X} \| \sigma_1) = H(\mathfrak{X} \| \sigma_2)$$

Proof. In the proof we shall use the abbreviations

$$\nu_1 = \nu_1(t) \qquad \xi_1 = \xi(1,\nu_1(t))$$

$$\nu_2^r = \nu_2^r(t) \qquad \xi_2^r = \xi(1,\nu_2^r(t)).$$

As ν_1 (ν_2^r) is uniquely determined by ξ_1 (ξ_2^r), we have

$$H(\xi_1|\xi_2^r) = H(\xi_1,\nu_1 \mid \xi_2^r,\nu_2^r). \tag{3.34}$$

We may rewrite the right hand side by (2.20)

$$H(\xi_1,\nu_1 \mid \xi_2^r,\nu_2^r) = H(\nu_1|\nu_2^r,\xi_2^r) + H(\xi_1|\nu_1,\nu_2^r,\xi_2^r). \tag{3.35}$$

Using the inequality (2.26) we get

$$H(\nu_1|\nu_2^r,\xi_2^r) \leq H(\nu_1|\nu_2^r). \tag{3.36}$$

As for given ν_1,ν_2^r and ξ_2^r the number of possible "values" of ξ_1 is at most $n^{|\nu_1-\nu_2^r|}$, for the last term in (3.35) $H(\xi_1|\nu_1=k_1,\nu_2^r=k_2,\xi_2^r=u) \leq \log n^{|\nu_1-\nu_2^r|}$ holds. Hence

$$H(\xi_1|\nu_1,\nu_2^r,\xi_2^r) = E\,H(\xi_1|\nu_1=k_1,\nu_2^r=k_2,\xi_2^r=u) \leq$$

$$\leq E\log n^{|\nu_1-\nu_2^r|} = E|\nu_1-\nu_2^r|\cdot\log n \tag{3.37}$$

For the right hand side of (3.36) the inequality

$$H(\nu_1|\nu_2^r) = H(\nu_1-\nu_2^r \mid \nu_2^r) \leq H(\nu_1-\nu_2^r) \tag{3.38}$$

holds. Apply the inequality (1.13)

$$\sum p_k \log \frac{p_k}{q_k} \geq 0$$

for $p_k = P(\nu_1 - \nu_2^r = k)$ and $q_k = \frac{1}{3} 2^{-|k|}$:

(3.39) $$H(\nu_1 - \nu_2^r) \le E|\nu_1 - \nu_2^r| + \log 3 .$$

Summarizing the results from (3.34) to (3.39) we obtain

(3.40) $$H(\xi_1|\xi_2^r) \le (1 + \log n) E|\nu_1 - \nu_2^r| + \log 3 .$$

The same is true if we change ν_1, ξ_1 and ν_2^r, ξ_2^r ;

(3.41) $$H(\xi_2^r|\xi_1) \le (1 + \log n) E|\nu_2^r - \nu_1| + \log 3 .$$

The relation (which is a consequence of (2.19))

$$H(\xi_1) + H(\xi_2^r|\xi_1) = H(\xi_2^r) + H(\xi_1|\xi_2^r)$$

used earlier will be usefull again. It results

$$|H(\xi_1) - H(\xi_2^r)| \le H(\xi_2^r|\xi_1) + H(\xi_1|\xi_2^r)$$

and applying (3.40) and (3.41)

$$|H(\xi_1) - H(\xi_2)| \le 2(1 + \log n) E|\nu_2^r - \nu_1| + 2\log 3 .$$

The statement of the theorem follows from (3.32) :

$$H(X\|\sigma_1) = \lim_{t\to\infty} \frac{H(\xi_1(1,\nu_1(t)))}{t} = \lim_{t\to\infty} \frac{H(\xi_2(1,\nu_2^r(t)))}{t} =$$

$$= \lim_{t\to\infty} \frac{1}{r} \frac{H(\xi_2(1,\nu_2(\frac{t}{r})))}{\frac{t}{r}} = \frac{1}{r} H(X\|\sigma_2) .$$

In order to apply Theorem 8 to concrete problems it will be convenient to establish some simple sufficient conditions of (3.32) for different cost-scales. We say that $\frac{\nu(t)}{t}$ is <u>uniformly integrable</u> if

$$\lim_{t\to\infty} \int_{\left|\frac{\nu(t)}{t}\right|\geq K} \left|\frac{\nu(t)}{t}\right| P(d\omega) \longrightarrow 0 \qquad \text{as } K\to\infty.$$

If $b \leq s(ux_{i_{k+1}})-s(u)$, where $b>0$ is a fixed constant then $\nu(t) \leq \frac{t}{b}$ results $\left|\frac{\nu(t)}{t}\right|$ is bounded, consequently it is uniformly integrable. However, the example of cost-scale defined by a block code shows that $s(ux_{i_{k+1}})-s(u)$ can be zero.

<u>Lemma 4. Let σ_1 and σ_2 be two cost scales having the properties</u>

$$b \leq s_2(ux_{i_{k+1}}) - s_2(u) \leq B \tag{3.42}$$

$(0 < b < B$ are constants $u = x_{i_1}\ldots x_{i_k}$, $k = 0,1,\ldots)$ and

$$\frac{\nu_1(t)}{t} \tag{3.43}$$

<u>is uniformly integrable then each of the conditions</u>

$$\frac{\nu_1(t)}{\nu_2\left(\frac{t}{r}\right)} \Longrightarrow 1 \tag{3.44}$$

$$\frac{s_2(\xi(1,\nu_1(t)))}{t} \Longrightarrow \frac{1}{r} \quad (\infty > r > 0) \tag{3.45}$$

$$\frac{s_1(\xi(1,\nu_2(t)))}{t} \Longrightarrow r \quad (\infty > r > 0) \tag{3.46}$$

<u>is equivalent to</u>

$$\frac{E\left|\nu_1(t) - \nu_2\left(\frac{t}{r}\right)\right|}{t} \Longrightarrow 0. \tag{3.47}$$

<u>Proof</u>. $\frac{\nu_1(t)}{t}$ and $\frac{\nu_2(t)}{t}$, thus $\frac{\nu_1(t)-\nu_2\left(\frac{t}{r}\right)}{t}$

are uniformly integrable. In this case (3.47) and

(3.48) $$\frac{\nu_1(t) - \nu_2\left(\frac{t}{r}\right)}{t} \Longrightarrow 0$$

are equivalent. (3.48) means

$$P\left(\left|\frac{\nu_1(t)-\nu_2\left(\frac{t}{r}\right)}{t}\right| > \varepsilon\right) = P\left(\left|\frac{\nu_1(t)-\nu_2\left(\frac{t}{r}\right)}{\nu_2\left(\frac{t}{r}\right)}\right| > \varepsilon\frac{t}{\nu_2\left(\frac{t}{r}\right)}\right) \to 0$$

and this is equivalent to (3.44) since $\left[\frac{t}{rB}\right] \le \nu_2\left(\frac{t}{r}\right) \le \frac{t}{rb}$ that is $rb \le \frac{t}{\nu_2\left(\frac{t}{r}\right)} \le 2rB$. Furthermore, as by definition of s_i's and ν_i's the relation $s_2(1,\xi(1,\nu_1(t))) \le yt$ is equivalent to $\nu_1(t) \le \nu_2(yt)$ $(0<y<\infty)$, the relation (3.45) is equivalent to

$$P(\nu_1(t) \le \nu_2(yt)) \longrightarrow \begin{cases} 0 & \text{if } y < \frac{1}{r} \\ 1 & \text{if } y < \frac{1}{r} \end{cases} \quad (t \to \infty)$$

and this, in turn, is equivalent to (3.48), in view of the assumption $0 < b \le s(ux_{i_{K+1}}) - s(u) < B$. Similarly, (3.46) is equivalent to

$$P(\nu_2(t) \le \nu_1(yt)) \longrightarrow \begin{cases} 0 & \text{if } y < r \\ 1 & \text{if } y > r \end{cases} \quad (t \to \infty)$$

i. e. to

$$P(\nu_2(y't) \le \nu_1(t)) \longrightarrow \begin{cases} 0 & \text{if } y' = \frac{1}{y} > \frac{1}{r} \\ 1 & \text{if } y' < \frac{1}{r} \end{cases} \quad (t \to \infty)$$

which, again, is equivalent to (3.48).

After Lemma 4 we may see that Lemma 3 is a particular case of Theorem 8, indeed. The first cost scale σ_1 is the counting-scale γ, the second one is the cost-scale defined by the given simple letter code. Both of them satisfy (3.42), consequently $\frac{\gamma(t)}{t}$ and $\frac{\nu_2(t)}{t}$ are uniformly integrable. Hence, by Lemma 4 the condition (3.32) of Theorem 8 is equivalent to (3.45):

$$\lim_{t\to\infty}\frac{s_2(\xi(1,\gamma(t)))}{t}=\lim_{t\to\infty}\frac{s_2(\xi(1,[t]))}{t}=\lim_{N\to\infty}\frac{\ell(\xi(1,N))}{N}\Longrightarrow\frac{1}{r}=L$$

and this is the condition of existing of the average code length in Lemma 3. The conditions result

$$H(X)=H(X\parallel\gamma)=LH(X\parallel\sigma_2)$$

in both cases.

4. More General Channels, more General Codes.

It is needless to say that the channels considered untill now are very special. First of all, we assumed that each code signal may follow each other. In the case of practical channels there are several conditions what sort of sequences of code signals are transmissible in the channel. For example we may exclude the sequences when y_i is followed by the same y_i.

The second particular property of the channels considered untill now is that each of the code signals has the same duration. The Morse-alphabet shows that in general it is not true. The 3 Morse-signals have three different durations. The general definition of the noiseless channel is the following.

The set of code signals is $Y=\{y_1,\dots,y_n\}$. Y^* and Y^∞ denotes the set of finite and infinite sequences formed from y_i's, respectively. $Y_T \subset Y^*$ is the set of <u>transmissible sequences</u> if it satisfies the following conditions :

(4.49) the void sequence $v_0 \in Y_T$,

(4.50) if $v \prec v'$ (v is a segment of v') and $v' \in Y_T$ then $v \in Y_T$

(4.51) if $v \in Y_T$ then there is a y_i such that $v y_i \in Y_T$.

On the other hand, a cost-function $f(v)$ is defined on Y_T $(v \in Y_T)$ which has three properties, too:

(4.52) $f(v_0) = 0$

(4.53) if $v \prec v'$ then $f(v) \leq f(v')$

(4.54) if $v_1 \prec \dots \prec v_k \prec \dots$ but $v_j \neq v_{j+1}$ $(1 \leq j)$ then $\lim\limits_{k\to\infty} f(v_k)=\infty$. Y_T and the cost-function f define a <u>noiseless channel</u> (Y_T, f). If $Y_T=Y^*$ the channel is called <u>memoryless channel</u>.

<u>Example 4</u>. <u>A finite state channel</u> has some states. In a fixed state it is able to transmit only a subset of code signals. After transmission of a signal the channel changes its state. The new state is defined by the last state and by the last transmit-

ted signal. The cost of a signal depends even on the present state of the channel. There is a fixed initial state before transmitting the first signal.

We give now the more formal form of this definition. Let $Y=\{y_1,\dots,y_m\}$ be the set of code signals and let $A=\{a_1,\dots,a_r\}$ be the set of states. The initial state is $a_i \in A$. For each $a_j \in A$ there is a subset $\mathfrak{I}(j) \subset \subset \{1,\dots,m\}$ of indices such that $Y(a_j)=\{y_k\}_{k\in\mathfrak{I}(j)}$ is the set of transmissible signals at the state a_j. The function $F(j,k)$ $(1 \leq j \leq r;\ k \in \mathfrak{I}(j))$ specifies the next state $a_{F(j,k)}$ if the signal y_k has been transmitted at the state a_j. A sequence $v=y_{j_1}\dots y_{j_N}$ is <u>transmissible</u> by the channel with initial state $a_i \in A$ $(v=y_{j_1}\dots y_{j_N} \in Y_T^i)$ if there is a sequence $i_0=i, i_1,\dots,i_N$ such that

$$j_k \in \mathfrak{I}(i_{k-1}) \qquad k=1,\dots,N \tag{c_1}$$

$$i_k = F(i_{k-1}, j_k) \qquad k=1,\dots,N. \tag{c_2}$$

The states of the channel are $a_i, a_{i_1}, \dots, a_{i_N}$; a_{i_k} is the state after having transmitted the signal y_{j_k}.

For each pair $a_j \in A$, $y_k \in Y(a_j)$ there is a fixed cost $f_{jk} \geq 0$. The cost of a sequence $v=y_{j_1}\dots y_{j_N}$ is

$$f(v)=\sum_{k=1} f_{i_{k-1} j_k}$$

where i_{k-1} $(1 \leq k \leq N)$ is defined by (c_1) and (c_2).

Let us define

$$f_j^* = \min_{k \in \mathcal{J}(j)} f_{j,k}.$$

There is one more technical condition. If $y_{j_1} \dots y_{j_N} \in Y_T^i$ and $N \geq r$, then there exist at least one $k \leq N$ for which $f_{i_k}^* > 0$. This condition ensures the finiteness of the capacity.

The channel is called indecomposable if for each pair of states a_i, a_k there exists a sequence $y_{j_1} \dots y_{j_N} \in Y_T^i$ for which $i_N = k$ (see (c_1) and (c_2)).

The notion of the finite-state channel is due to Shannon and Weaver [4].

Untill now we have used two types of codes : simple letter codes and block codes. However, theoretically there are many possibilities to make codes. For instance, we may change the lengths of blocks according to a fixed rule. The next example shows a more unusual code.

Example 5. Put $X = \{x_1, x_2\}$ and $Y = \{x_1, 0, 1\}$. Before encoding we divide the sequence of information signals into blocks. x_1 forms a block in itself. The sequences of x_2's without gaps form blocks, too. Now we encode block by block. $c(x_1) = x_1$ and $c(x_2 \dots x_2)$ is the binary form of the number of x_2's, written by 0's and 1's.

Assume the channel is memoryless and the cost is unit. If the probability p_1 of x_1 is very small then the above code seems to be very good. Is it possible that the average code length is less than $\frac{H(X)}{\log 3}$? Our results up to this point do not contradict this possibility. But we shall solve it soon.

We give now a general definition of the code.

Let $X = \{x_1, \ldots, x_N\}$ be the set of information signals. X^* denotes the set of finite sequences formed from the elements of X. A code is a function $c(u)$ $(u \in X^*)$ which maps X^* into Y^* and satisfies the following conditions :

(4.55) $c(u_0) = v_0$

(4.56) $u \prec u'$ implies $c(u) \prec c(u')$

(4.57) if $u_1 \prec \ldots \prec u_k \prec$ but $u_j \neq_{j+1}$ $(1 \leq j)$

then $\lim_{k \to \infty} |c(u_k)| = \infty$ where $|v|$ means the number of elements of the sequence v.

The code $c(u)$ is admissible for a given channel (Y_T, f) if it maps X^* into Y_T.

Finally we have to give one more definition. A code will be said to be finite decodable if there is a natural number d such that to any fixed v there are at most d u's satisfying $c(u) = v$.

The question arises, is not it sufficient to use the notion for $d = 1$. The example of an N-block code answers no ! In this case $c(u) = v_0$ if $|u| < N$, thus $d = \sum_{i=0}^{N-1} m^i = \frac{m^N - 1}{m - 1}$. Example 5 gives a code which is not finite decodable. A more general definition of decodability is needed which includes Example 5. This notion of decodability, unfortunately, depends even on the encoded source $\mathfrak{X}$. We say that a code $c(u)$ defined on the information source $\mathfrak{X}$ is <u>quasi-finite decodable</u> if there exists a real number e such that for any fixed v

$$- \sum_{u \in X^*} P(u|v) \log P(u|v) \le e$$

holds. If the code is finite decodable than $e = \log d$. The code in Example 5 is quasi-finite decodable (assume the information source emits independent signals). Namely, it is easy to see that

$$- \sum_{u \in X^*} P(u|v) \log P(u|v) \le \sum_{i=0}^{\infty} p_1 p_2^i \log p_1 p_2^i = p_1 \log p_1 \sum_{i=0}^{\infty} p_2^i + p_1 \log p_2 \sum_{i=0}^{\infty} i p_2^i,$$

and this is a finite number if $p_2 < 1$.

Now we are able to state "the principle of conservation of entropy" in a general form.

<u>Theorem 9. Let $\mathfrak{X}$ be an information source with a given cost-scale σ_1, where $H(\mathfrak{X} \| \sigma_1)$ exists. Let the code c, defined on $\mathfrak{X}$, be a quasi finite</u>

decodable code which is admissible for the channel (Y_T, f) and let σ_3 be the cost-scale specified by $\hat{f} = s_3$ for the source $\mathcal{Y}$ defined by $\mathcal{X}$ and c. If the assumptions

$$\frac{\nu_1(t)}{t} \text{ is uniformly integrable if } t \to \infty , \tag{4.58}$$

$$b \leq s_3(v y_i) - s_3(v) \leq B , \quad (0 < b < B, 1 \leq i \leq m, v \in Y_T), \tag{4.59}$$

$$\frac{s_3(c(\xi(1,\nu_1(t))))}{t} \Longrightarrow L > 0 , \tag{4.60}$$

$$\frac{|c(u)|}{|u|} \text{ is bounded} , \tag{4.61}$$

hold, then

$$H(\mathcal{Y} \| \sigma_3) = \frac{H(\mathcal{X} \| \sigma_1)}{L} . \tag{4.62}$$

Proof. We define a cost-scale σ_2 on $\mathcal{Y}$ by

$$\nu_2(t) = |c(\xi(1,\nu_1(t)))| . \tag{4.63}$$

hence

$$\eta(1,\nu_2(t)) = c(\xi(1,\nu_1(t))) \tag{4.64}$$

follows where $\eta(1,N)$ denotes the random vector of the first N signal of the encoded sequence. The relation

$$H(c(\xi(1,\nu_1(t))) \mid \xi(1,\nu_1(t))) = 0 \tag{4.65}$$

is obvious. On the other hand, by the quasi-finite decodability

$$H(\xi(1,\nu_1(t)) \mid c(\xi(1,\nu_1(t)))) \le e . \tag{4.66}$$

Using

$$H(\xi(1,\nu_1(t))) + H(c(\xi(1,\nu_1(t))) \mid \xi(1,\nu_1(t))) =$$

$$= H(c(\xi(1,\nu_1(t)))) + H(\xi(1,\nu_1(t)) \mid c(\xi(1,\nu_1(t))))$$

we get from (4.65) and (4.66)

$$H(\mathcal{X}\|\sigma_1) = \lim_{t\to\infty} \frac{H(\xi(1,\nu_1(t)))}{t} = \lim_{t\to\infty} \frac{H(c\ \xi(1,\nu_1(t)))}{t} =$$

$$= \lim_{t\to\infty} \frac{H(\eta(1,\nu_2(t)))}{t} = H(\mathcal{Y}\|\sigma_2) \tag{4.67}$$

If $\frac{\nu_2(t)}{t}$ then by (4.63)

$$\frac{\nu_2(t)}{t} = \frac{|c(\xi(1,\nu_1(t)))|}{t} \le K \frac{|\xi(1,\nu_1(t))|}{t} = K \frac{\nu_1(t)}{t}$$

that is, $\frac{\nu_2(t)}{t}$ is uniformly integrable, too.

We may apply Lemma 4 for $\mathcal{Y}$, σ_2 and σ_3. In our case (4.45) has the form $\frac{s_3(\eta(1,\nu_2(t)))}{t} \Longrightarrow \frac{1}{L} = L > 0$ and this is equivalent to (4.60) by (4.64). From the lemma we obtain

$$\frac{E|\nu_2(t) - \nu_3(Lt)|}{t} \to 0 .$$

All the conditions of Theorem 8 hold, we apply it for $\mathcal{Y}, \sigma_2$ and σ_3 :

$$H(\mathcal{Y}\|\sigma_2) = L H(\mathcal{Y}\|\sigma_3) . \tag{4.68}$$

The desired equation (4.62) follows from (4.67) and (4.68).

In pratical cases we want to minimize L. After Theorem 9 we get an upper bound for L maximizing $H(\mathcal{Y}\|\sigma_3)$ in (4.62). We did the same thing after Theorem 6 ; Theorem 2 was a simple consequence of it. Let $N(t)$ denote the number of possible values of $\eta(1,\nu_3(t))$ that is "number of possible different events untill t". By the definition of $H(\mathcal{Y}\|\sigma_3)$ we get

$$H(\mathcal{Y}\|\sigma_3) = \lim_{t\to\infty} \frac{H(\eta(1,\nu_3(t)))}{t} \leq \overline{\lim_{t\to\infty}} \frac{\log N(t)}{t} = C. \quad (4.69)$$

The number C is the capacity of the channel $(\mathcal{Y}_T, \hat{r})$ $(\hat{r} = s_3)$. (It does not depend on the probability distributions of $\mathcal{Y}$).

Hence we obtain the noiseless coding theorem which is a simple consequence of Theorem 9, however it is a relatively new result in this generality. It expresses that a channel is not able to transmit more information within the unit time than its capacity.

Theorem 10. Under the conditions of Theorem 9

$$\frac{H(\mathcal{X}\|\sigma_1)}{C} \leq L. \quad (4.70)$$

Proof. The theorem follows from Theorem 9 and (4.69).

The definition and theorems of this par-

agraph are formulated in more general form in [3]. As regards to (4.70) for the case $\sigma_1 = \sigma_2 = \gamma$ [12] contains a stronger result.

In the remaining part of these lecture notes we shall determine the capacity C in particular cases and we shall examine the problem, when can we approximate the lower bound (4.70) (see Theorem 4 earlier).

5. The Capacity of a Memoryless Channel.

Recall that a channel (Y_T, f) is memoryless if $Y_T = Y^*$. We restrict ourselves for the case when the costfunction f has the form

$$f(v) = f(y_{i_1} \dots y_{i_k}) = \sum_{j=1}^{k} f_{i_j}$$

where f_i is the cost of the signal y_i. We assume

$$f^{**} = \max(f_1, \dots, f_m) > \min(f_1, \dots, f_m) = f^* > 0.$$

As earlier, $N(t)$ denotes the number of different sequences $v = y_{i_1} \dots y_{i_k}$ such that

(5.71) $$\sum_{j=1}^{k} f_{i_j} \leq t$$

however for some i

$$\sum_{j=1}^{k} f_{i_j} + f_i > t\,.$$

On the other hand $N'(t)$ denotes the number of sequences

(5.72) $v = y_{i_1} \dots y_{i_k}$ such that $\sum_{j=1}^{k} f_{i_j} \leq t$ but $\sum_{j=1}^{k} f_{i_j} + f^* > t$.

If a sequence v satisfies (5.72) then it satisfies also (5.71). Hence

$$N'(t) \le N(t). \qquad (5.73)$$

In turn, from a sequence satisfying (5.72) we may omit the last elements untill we get a sequence satisfying (5.71). We reach in this manner all the sequences satisfying (5.71). How many is reached from one sequence ? At most $\frac{f^{**}}{f^{*}}$, that is, the inequality

$$N(t) \le \frac{f^{**}}{f^{*}} N'(t) \qquad (5.74)$$

holds. By (5.73) and (5.74) we get

$$C = \lim_{t\to\infty} \frac{\log N(t)}{t} = \lim_{t\to\infty} \frac{\log N'(t)}{t}$$

which means we may use $N'(t)$ instead of $N(t)$ to determine the capacity.

Let us consider the function

$$a(\omega) = \sum_{i=1}^{m} \omega^{-f_i}.$$

It is a continuous and monotonically decreasing function in the interval $(0,\infty)$. Since $a(1) \ge 2$ $(m \ge 2)$, the equation $a(\omega)=1$ has a unique positive root $\omega_0 > 1$.

<u>Lemma 5. There is a positive constant</u> b <u>such that</u>

$$b\,\omega_0^t \le N'(t) \le \omega_0^t \qquad (t \ge 0) \qquad (5.75)$$

<u>where ω_0 is the unique positive root of</u> $a(\omega) = 1$.

<u>Proof</u>. Put $N(t)=0$ if $t<0$. If $0 \le t < f^*$ then trivially $N(t)=1$. If $f^* \le t$ the number of sequences satisfying (5.72) and beginning with y_i is obviously $N'(t-f_i)$. Hence

(5.76) $$N'(t) = \sum_{i=1}^{m} N'(t-f_i).$$

Define the positive numbers b_k recursively by

$$b_1 = \omega_0^{-f^*}, \quad b_{k+1} = b_k \sum_{f_i \le kf^*} \omega_0^{-f_i} \quad (k=1,2,\dots).$$

We prove the inequality

(5.77) $$b_k \omega_0^t \le N'(t) \le \omega_0^t \qquad (0 \le t < kf^*)$$

by induction over k. (5.77) clearly holds for $k=1$; moreover, if it holds for some positive integer k, for $kf^* \le t < (k+1)f^*$ we have, in view of the assumption

$$b_{k+1}\omega_0^t \le \sum_{f_j \le t} b_k \omega_0^{t-f_j} \le \sum_{j=1}^{m} N'(t-f_j) \le \sum_{j=1}^{n} \omega_0^{t-f_j} = \omega_0^t .$$

Thus, on account of (5.76),(5.77) holds for $k+1$ if it holds for k. (Observe that the sequence b_k is non-decreasing).

To complete the proof of (5.75) we have only to notice that $b = \lim_{k\to\infty} b_k$ is positive and $b_k \ge b\,(1 \le k)$.

<u>Theorem 11. The capacity of a memoryless</u>

channel is

$$C = \log \omega_0$$

where ω_0 is the unique positive root of $a(\omega)=\sum_{j=1}^{m}\omega^{-f_j}=1$.

Proof. It is a trivial consequence of Lemma 5.

The theorem was formulated first by Shannon and Weaver [4],nevertheless, the rigorous proof has been given by Krause [5] (in a little bit different variant). This simplest proof is due to Csiszár [6].

6. Finite-State Channels.

In Example 4 we defined the finite state channels. Now, we want to determine their capacity. Let $N_i'(t)$ denote the number of sequences $v = y_{j_1}\dots y_{j_N} \in Y_T^i$ satisfying

$$\sum_{k=1}^{N} f_{i_{k-1}j_k} \le t\ ,\quad \sum_{k=1}^{N} f_{i_{k-1}j_k} + f_{i_k}^* > t, \qquad (6.78)$$

where $f_j^* = \min_{k\in\mathcal{F}(j)} f_{jk}$. Where as, $N_i(t)$ means the number of sequences $v = y_{j_1}\dots y_{j_N} \in Y_T^i$ satisfying $\sum_{k=1}^{N} f_{i_{k-1}j_k} \le t$, but for some $y_{j_{N+1}} \in Y(a_N): \sum_{k=1}^{N} f_{i_{k-1}j_k} + f_{i_N j_{N+1}} > t$. $N_i'(t)$ and $N_i(t)$ are different in general. However, we may prove

$$C_i = \lim_{t\to\infty} \frac{\log N_i(t)}{t} = \lim \frac{\log N_i'(t)}{t}$$

(in the sense if one of them exists then the second also exists) by the same arguments as in the preceding paragraph.

Let $\mathcal{A}(\omega)$ denote the $r \times r$ matrix

$$\mathcal{A}(\omega) = (a_{ik}(\omega)) \qquad a_{ik}(\omega) = \sum_{j \in \mathcal{J}_k(i)} \omega^{-f_{ij}}$$

where

$$\mathcal{J}_k(i) = \{ j : j \in \mathcal{J}(i),\ F(i,j) = k \},$$

that is, the set of indices j which are transmissible at state a_i and for which y_j changes the state of the channel from a_i into a_k.

Lemma 6. Let us be given an indecomposable finite-state channel. Then there exist positive numbers $0<b<B$ such that

$$b\,\omega_0^t \leq N_i'(t) \leq B\,\omega_0^t \qquad (i=1,\dots,r) \tag{6.79}$$

where ω_0 is defined as the (unique) positive number ω for which the greatest positive eigenvalue of the matrix $\mathcal{A}(\omega)$ equals one. This ω_0 is the greatest positive root of the equation

$$\operatorname{Det}\Big(\sum_{j \in \mathcal{J}_k(i)} \omega^{-f_{ij}} - \delta_{ik} \Big) = 0 \tag{6.80}$$

where δ_{ik} is Kronecker's delta.

Proof. Put $N_i'(t) = 0$ if $t<0$. It is

obvious that $N_i'(t)=1$ if $0\leq t<f_i^*$. If $f_i^*\leq t$, the number of sequences satisfying (6.78) and beginning with $y_j(j\in\mathfrak{Z}(i))$ is equal to $N'_{F(i,j)}(t-f_{ij})$. Hence, by definition of $\mathfrak{Z}_k(i)$

$$N_i'(t)=\sum_{k=1}^{r}\sum_{j\in\mathfrak{Z}_k(i)}N_k'(t-f_{ij})\quad\text{if}\quad f_i^*\leq t\quad(1\leq i\leq r). \tag{6.81}$$

As $\mathcal{A}(\omega)$ is a matrix with non-negative elements, its greatest positive eigenvalue $\lambda(\omega)$ is equal to the least upper bound of the set of positive λ's satisfying $\sum_{k=1}^{r}a_{ik}(\omega)b_k\geq\lambda b_i$ $(1\leq i\leq r)$ for some r-tuple of non-negative numbers $b_1,\ldots,b_r$ not all equal to zero Moreover, as the matrix is indecomposable (this is obviously equivalent to the indecomposability of the channel), the components of the eigenvector of $\mathcal{A}(\omega)$ belonging to its greatest positive eigenvalue are positive.

The above representation of $\lambda(\omega)$ implies that $\lambda(\omega)$ is a strictly decreasing, continuous function of (ω), with $\lambda(1)\geq1$ and $\lim_{\omega\to\infty}\lambda(\omega)=0$. Thus there exists a unique positive number $\omega_0\geq1$ with $\lambda(\omega_0)=1$. Let the components of eigenvector of $\mathcal{A}(\omega)$ belonging to the eigenvalue $\lambda(\omega_0)=1$ be denoted by b_i $(1\leq i\leq r)$.

---

* Concerning the simple properties of matrices with non-negative elements used below, see e. g. [7].

Then $b_i > 0$ $(1 \le i \le r)$ and

$$(6.82) \quad \sum_{k=1}^{r} \sum_{j \in \mathcal{J}_k(i)} b_k \omega_0^{-f_{ij}} = \sum_{k=1}^{r} b_k a_{ik}(\omega_0) = b_i \quad (1 \le i \le r).$$

Let now d and D be positive numbers such that in the interval $0 \le t < f^{**} = \max\limits_{\substack{1 \le i \le r \\ j \in \mathcal{J}(i)}} f_{ij}$

$$(6.83) \qquad d b_i \omega_0^t \le N_i'(t) \le D b_i \omega_0^t \quad (1 \le i \le r).$$

As the numbers b_i are all positive and f^{**} is finite, such positive numbers $d < D$ surely exist. Then one verifies in the same way as in the proof of Lemma 5 that (6.83) holds for every $t \ge 0$. In fact, if

$$\hat{f} = \min_{\substack{1 \le i \le r \\ j \in \mathcal{J}(i)}} f_{ij}$$

and (6.83) is valid for $0 \le t \le f^{**} + N\hat{f}$ then for $f^{**} + N\hat{f} \le t < f^{**} + (N+1)\hat{f}$ we have $0 \le t - f_{ij} < f^{**} + N\hat{f}$ $(1 \le i \le r, j \in \mathcal{J}(i))$ and (6.82) implies

$$d b_i \omega_0^t = \sum_{k=1} \sum_{j \in \mathcal{J}_k(i)} d b_k \omega_0^{t - f_{ij}} \le \sum_{k=1} \sum_{j \in \mathcal{J}_k(i)} N_k'(t - f_{ij}) \le$$
$$\le \sum_{k=1} \sum_{j \in \mathcal{J}_k(i)} D b_k \omega_0^{t - f_{ij}} = D b_i \omega_0^t ;$$

this means, on account of (6.81), that (6.83) holds for $0 \le t < f^{**} + (N+1)\hat{f}$ as well.

If $f_{ij} = 0$ for some pair (i, j) i. e. if $\hat{f} = 0$, the above induction breaks down. However, in view of the last assumption of the definition of the finite-state channel, one has for every finite-state channel

$$f = \min_{\substack{1 \le i \le r \\ y_{j_1} \dots y_{j_r} \in Y_T^i}} \sum_{k=1} t_{i_{k-1} j} > 0 \ .$$

Then the above argument can be modified by letting $\tilde{f}$ and rf^{**} play the role of $\hat{f}$ and f^{**}, respectively, and referring to the r's iterates of systems of equations (6.81) and (6.82) (instead of (6.81) and (6.82) themselves).

Thus for any indecomposable finite-state channel the inequalities (6.83) hold for every proving (6.79).

Obviously, ω_0 is a positive root of equation (6.80). Moreover, as the greatest positive eigenvalue $\lambda(\omega)$ of $\mathcal{A}(\omega)$ is strictly decreasing function of ω, in the case $\omega > \omega_0$ the number λ cannot be an eigenvalue of $\mathcal{A}(\omega)$ thus ω cannot be a root of (6.80) ; i. e. ω_0 is the greatest positive root of equation (6.80).

The following theorem is a trivial consequence of the above lemma.

<u>Theorem 12. If an indecomposable finite-state channel is given, then its capacity is</u>

$$C = C_i = \log_2 \omega_0 ,$$

<u>where ω_0 is the greatest positive root of the equation</u>

(6.80).

For this theorem the first rigorous proof was given by Ljubic [8] . This simple proof is due to Csiszár [6] .

Finally, let us describe an encoding procedure for finite-state channel which approximates the channel capacity. Let $X=\{x_1,\dots,x_n\}$ be the set of information signals, and let $\mathcal{X}$ be an information source which emits signals from X independently with probabilities $p_1 \geq \dots \geq p_n$. We make correspond to each $x_i \in X$ a code word $c_k(x_i) = y_{j_1} \dots y_{j_n}$ (depending on k $(1 \leq k \leq r)$) in the following way. Put

(6.84) $$\alpha_i = \sum_{\ell=1}^{i-1} p_\ell \qquad (1 \leq i \leq n).$$

Let us subdivide the unit interval into $|\mathfrak{J}(i_0)|$ disjoint (left semi-closed) subintervals I_{j_1}, $j_1 \in \mathfrak{J}(i_0)$ of length $(1/b_{i_0}) b_{F(i_0,j_1)} \omega_0^{-f_{i_0 j_1}}$ where the numbers b_k and ω_0 are the same as in the proof of Lemma 6. In the next step let us subdivide each I_{j_1} $(j_1 \in \mathfrak{J}(i_0))$ into $|\mathfrak{J}(i_1)|$ $(i = F(i_0, j_1))$ disjoint subintervals $I_{j_1 j_2}$ $(j_2 \in \mathfrak{J}(i_1))$ of length $(1/b_{i_0})\, b_{F(i_1,j_2)} \cdot \omega_0^{-(f_{i_0 j_1} + f_{i_1 j_2})}$, an so on ; in the N th step let us subdivide each $I_{j_1 \dots j_{N-1}}$ into $|\mathfrak{J}(i_{N-1})|$ $(i_{N-1} = F(i_{N-2}, j_{N-1}))$

disjoint subintervals $I_{j_1 \dots j_N}$ $(j_N \in \mathcal{J}(i_{N-1}))$ of length $(1/b_{i_0})\, b_{F(i_{N-1}, j_N)}\, \omega^{-\sum_{l=1}^{N} f_{i_{l-1} j_l}}$. Now we choose (see (6.84)) $c_k(x_i) = y_{j_1} \dots y_{j_N}$ if $\alpha_i \in I_{j_1 \dots j_N}$ and neither α_{i+1} nor α_{i-1} belongs to $I_{j_1 \dots j_N}$ while either α_{i+1} or α_{i-1} (or both) belongs to $I_{j_1 \dots j_{N-1}}$.

Theorem 13. The encoding $x_i \to c_k(x_i)$ described above has the prefix property ; the code words $c_k(x_i)$ are transmissible by the channel if the initial state is $i_0 = k$ and the average code cost $L = \sum_{i=1}^{n} p_i \ell_i$ satisfies

$$L < \frac{H(\mathfrak{X}) + E}{C} + f^{**} \tag{6.85}$$

where

$$\ell_i = \sum_{k=1}^{N} f_{i_{k-1} j_k} \quad \text{if} \quad c_k(x_i) = y_{j_1 \dots j_N}. \tag{6.86}$$

E and f^{**} are constant depending on the channel only and $C = \log \omega_0$ is the channel capacity.

Proof. The prefix property of the encoding $x_i \to c_k(x_i)$ and the assertion that $c_k(x_i)$ is transmissible if the initial state is k i.e. that $c_k(x_i) \in Y_T^i$ follow from the construction of the $c_k(x_i)$'s (see C_1 and (C_2)) . It is also obvious from the construction that if $c_k(x_i) = y_{j_1} \dots y_{j_N}$ then the length of $I_{j_1 \dots j_{N-1}}$ is greater than

$$\min(\alpha_i - \alpha_{i-1}, \alpha_{i+1} - \alpha_i) = p_i.$$

This means

$$(6.87)\qquad \frac{1}{b_{i_0}}\, b_{i_{N-1}}\, \omega_0^{-\sum_{\ell=1}^{N-1} f_{i_{\ell-1} i_\ell}} > p_i \,.$$

Write $E' = \max_{\substack{1 \le i \le r \\ 1 \le k \le r}} (b_i / b_k)$ and $f^{**} = \max_{\substack{1 \le i \le r \\ j \in \mathcal{J}(i)}} f_{ij}$

then (6.86) and (6.87) imply

$$(\ell_i - f^{**}) \log \omega_0 < -\log p_i + \log E'$$

hence, on account of Theorem 12 and the definition of $H(\mathfrak{X})$ we obtain (6.85), with $E = \log E'$.

This encoding procedure is given by Csiszár [6] . This is a generalization of the procedure of Krause [5] (for memoryless channel with different costs) which is a generalization of the familiar Shannon-Fano code.

References.

[1] Karush, J. : A simple proof of an inequality of McMillan, IRE Trans. IT-7, 118.

[2] Katona, G. and Tusnády, G. : The principle of conservation of entropy in a noiseless channel, Studia Sci. Math. Hungar.2 (1967)29-35.

[3] Csiszár, I., Katona, G. and Tusnády, G. : Information sources with different cost scales and the principle of conservation of entropy, Z. Wahrscheinlichkeitstheorie und Verw. Gebiete 12 (1969) 185-222.

[4] Shannon, C.E. and Weaver, W. : The Mathematical Theory of Communication, University of Illinois Press, Urbana, Ill. 1949.

[5] Krause, R.M. : Channels which transmit letters of inequal duration, Information and Control 5(1962)13-24.

[6] Csiszár, I. : Simple proof of some theorems on noiseless channels, Information and Control 14 (1969) 285-298.

[7] Gantmacher, F.R. : Applications of the theory of Matrices , Interscience Publishers, New York, 1959.

[8] Ljuvic, Ju.I. : Remark on the capacity of a dis-

crete channel without noise (in Russian), Uspehi Mat. Nauk 17 (1962) 191-198.

[9] Feinstein, A. : Foundations of Information Theory, McGraw Hill, New York, 1958.

[10] Sidel'nikov, V.M. : On statistical properties of transformations induced by finite automata (in Russian), Kibernetika (Kiev) 6 (1965) 1-14.

[11] Shannon, C.E. : A mathematical theory of communication, Bell System Techn. J. 27(1948)379 379-432, 623-656.

[12] Billingsley, P. : On the coding theorem for the noiseless channel, Ann. Math. Statist. 32 (1961) 594-601.

Contents

Printed in the United States
By Bookmasters